# Memorizing
# the periodic table of chemical elements.

A simple method.

# Contents

## Introduction

*In the fall of 1997 I took an Introduction to Chemistry class (Chem 100) at Tacoma Community College, Tacoma, Washington. We were required to memorize at least eight or ten of the more important chemical elements. I asked myself at the time —*

**Is it possible to memorize the number of protons and their symbols for the entire periodic table? Is it also possible to include memorization of nuclear mass numbers? The answer to both questions is yes, and the process is a simple one as you will see.**

*I hope the information in this little book will be helpful and an encouragement to those of you who are studying chemistry. Maybe some of you can even surprise your teachers or professors with the results of your memory work. More importantly, I hope that the technique I have presented here will inspire confidence in those who find it of interest and stimulate an interest in the study of chemistry and science in general.*

*D.H.D.*

*Renton, Washington*
*16 November 2007*

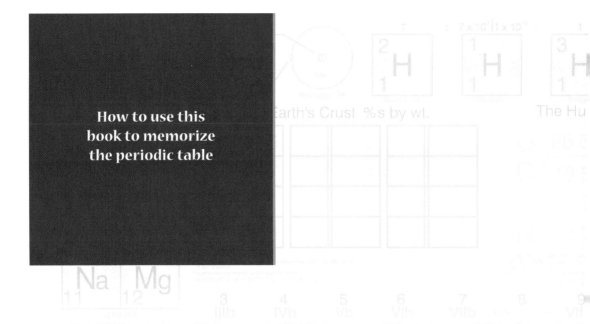

## The Number of Protons

■ The first part of the memorization process is to use a blank Periodic Table. This book provides one that you can duplicate for memory work. You can also construct one with two sheets of college-ruled paper, a straight edge, and a pen or pencil. Building one for yourself is a very helpful aid in memorization because it requires you to learn the number of chemical elements in each row and the number of rows. Either way, once you have a blank Periodic Table you are ready to begin your memory work.

First of all, a chemical element is identified by the number of protons. Hydrogen has one, Helium two, Lithium has three, etc. The number of protons goes in sequential order from 1 to 112, each number representing an individual chemical element. Memorization requires associating the number of protons with the chemical element and its symbol. Its physical place in the Periodic Table helps in memorization. This book provides a complete sequential listing of chemical elements by the number of protons (Table of Proton Numbers) to aid you in the memorization process.

## Nuclear Mass Numbers

■ The second part of the memorization process, once the symbols for the chemical elements and the number of protons have been recorded is to memorize the formulas and the series of black numbers shown in the Memorization Key. These will help you to easily derive the nuclear mass numbers.

Note that there are two simple algebraic formulae: $1X + $ (a number) and $2X + $ (a number). The 1X formula is used to derive the nuclear mass numbers for the three isotopes of Hydrogen (H) which are protium, deuterium, and tritium. The 2X formula is used for the remaining chemical elements in the Periodic Table. In both cases, X represents the number of protons for the element you are deriving the nuclear mass number for.

The nuclear mass number for protium is $1X + 0 = 1$. The nuclear mass number for deuterium is $1X + 1 = 2$. The nuclear mass number for tritium is $1X + 2 = 3$. These are the three isotopes of Hydrogen (H), each having the same number of protons (one), but differing in whether they have 0, 1, or 2 neutrons. We have the same chemical element, but different nuclear mass numbers.

The nuclear mass number for Helium (He) is $2X + 0 = 4$. Helium has 2 protons and 2 neutrons in its nucleus. The nuclear mass number for

anism %s by wt.

VIIIa 18

non-metals

| 13 IIIa | 14 IVa | 15 Va | 16 VIa | 17 VIIa | |
|---|---|---|---|---|---|
| 11 B 5 | 12 C 6 | 14 N 7 | 16 O 8 | 19 F 9 | 20 Ne 10 |
| 27 Al 13 | 28 Si 14 | 31 P 15 | 32 S 16 | 35 Cl 17 | 40 Ar 18 |

4 He 2

Iron (Fe) is 2X + 4 = 56. X here = 26. Iron (Fe) has 26 protons and 26 + 4 neutrons. Subtracting the number of protons from the nuclear mass number gives you the number of neutrons: 56 – 26 = 30. There are 30 neutrons in the nucleus of the chemical element Iron (Fe).

You will notice on the bottom of the Memorization Key, the number [50] - 1 followed by a series of numbers preceded by a plus or minus sign.

Beginning with Francium (Fr) the large, black numbers in the Memorization Key become more difficult to memorize. I used the nuclear mass number for Radon (Rn) -1 to give me the memorization number for Francium (Fr).

The memorization number for Radium (Ra) is the memorization number for Francium (Fr) +1, which is 50.

The memorization number for Thorium (Th) is the memorization number for Actinium (Ac) +3, which is 52. These numbers are much easier to memorize.

You may find it helpful to memorize the Memorization Key numbers by rows and to write those numbers down on a folded piece of paper from memory under the row you want to record the nuclear mass numbers on. You can do your addition there.

**1.** I have chosen the phrase *"the number of protons"* to replace the phrase *"the atomic numbers"* for philosophy of science reasons. The word atom in Greek (άτομοσ) means indivisible or uncut. All the chemical elements of the Periodic Table can be broken down into smaller parts.

**2.** This number and the others are rounded off for memory purposes. The full numbers are given in Table of Proton Numbers for your reference.

**3.** The number of electrons that are whirling about the nucleus of a chemical element is equal to the number of protons. Since Iron (Fe) has 26 protons, it has 26 electrons as a chemical element.

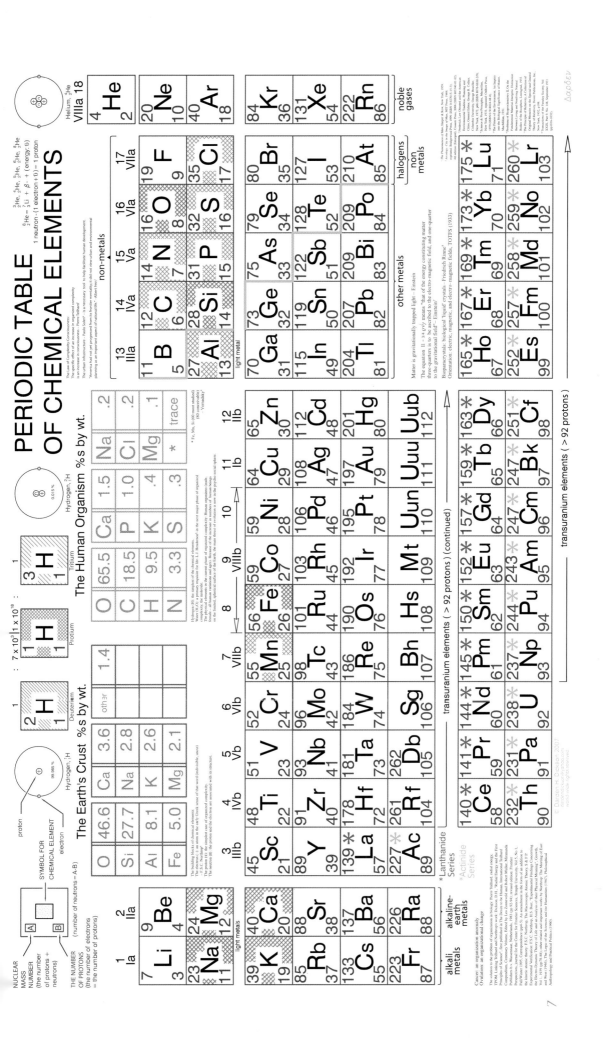

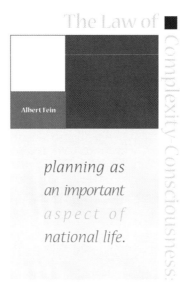

The Law of

Complexity Consciousness:

Albert Fein

*planning as*
*an important*
*aspect of*
*national life.*

# NUCLEAR MASS NUMBERS-MEMORIZATION KEY

**Legend**

| | |
|---|---|
| 2 / 1 / x+1 — Deuterium | 1 / 0 / x+0 — Protium |
| 3 / 2 / x+2 — Tritium | |

Where X = the number of protons

**Isotope cells (mass number / value / formula)**

| 7 / 1 / 2x+1 | 9 / 1 / 2x+1 |
|---|---|
| 23 / 1 / 2x+1 | 24 / 0 / 2x+0 |

| 4 / 0 / 2x+0 |
|---|
| 20 / 0 / 2x+0 |
| 40 / 4 / 2x+4 |

| 11 / 1 / 2x+1 | 12 / 0 / 2x+0 | 14 / 0 / 2x+0 | 16 / 0 / 2x+0 | 19 / 1 / 2x+1 |
|---|---|---|---|---|
| 27 / 1 / 2x+1 | 28 / 0 / 2x+0 | 31 / 1 / 2x+1 | 32 / 0 / 2x+0 | 35 / 1 / 2x+1 |

| 84 / 12 / 2x+12 |
|---|
| 131 / 23 / 2x+23 |
| 222 / 50 / 2x+50 |

| 80 / 10 / 2x+10 | 79 / 11 / 2x+11 | 75 / 9 / 2x+9 | 73 / 9 / 2x+9 | 70 / 8 / 2x+8 |
|---|---|---|---|---|
| 127 / 21 / 2x+21 | 128 / 24 / 2x+24 | 122 / 20 / 2x+20 | 119 / 19 / 2x+19 | 115 / 17 / 2x+17 |
| 210 / 40 / 2x+40 | 209 / 41 / 2x+41 | 209 / 43 / 2x+43 | 207 / 43 / 2x+43 | 204 / 42 / 2x+42 |

| 39 / 1 / 2x+1 | 40 / 0 / 2x+0 |
|---|---|
| 85 / 11 / 2x+11 | 88 / 12 / 2x+12 |
| 133 / 23 / 2x+23 | 137 / 25 / 2x+25 |
| 223 / 49 / 2x+49 | 226 / 50 / 2x+50 |

| 45 / 3 / 2x+3 | 48 / 4 / 2x+4 | 51 / 5 / 2x+5 | 52 / 4 / 2x+4 | 55 / 5 / 2x+5 | 56 / 4 / 2x+4 | 59 / 5 / 2x+5 | 59 / 5 / 2x+5 | 64 / 6 / 2x+6 | 65 / 5 / 2x+5 |
|---|---|---|---|---|---|---|---|---|---|
| 89 / 11 / 2x+11 | 91 / 11 / 2x+11 | 93 / 11 / 2x+11 | 96 / 12 / 2x+12 | 98 / 12 / 2x+12 | 101 / 13 / 2x+13 | 103 / 13 / 2x+13 | 106 / 14 / 2x+14 | 108 / 14 / 2x+14 | 112 / 16 / 2x+16 |
| 139* / 25 / 2x+25 | 178 / 34 / 2x+34 | 181 / 35 / 2x+35 | 184 / 36 / 2x+36 | 186 / 36 / 2x+36 | 190 / 38 / 2x+38 | 192 / 38 / 2x+38 | 195 / 39 / 2x+39 | 197 / 39 / 2x+39 | 201 / 41 / 2x+41 |
| 227* / 49 / 2x+49 | 261 / 53 / 2x+53 | 262 / 52 / 2x+52 | | | | | | | |

| 140* / 24 / 2x+24 | 141* / 23 / 2x+23 | 144* / 24 / 2x+24 | 145* / 23 / 2x+23 | 150* / 26 / 2x+26 | 152* / 26 / 2x+26 | 157* / 29 / 2x+29 | 159* / 29 / 2x+29 | 163* / 31 / 2x+31 | 165* / 31 / 2x+31 | 167* / 31 / 2x+31 | 169* / 31 / 2x+31 | 173* / 33 / 2x+33 | 175* / 33 / 2x+33 |
|---|---|---|---|---|---|---|---|---|---|---|---|---|---|
| 232* / 52 / 2x+52 | 231* / 49 / 2x+49 | 238* / 54 / 2x+54 | 237* / 51 / 2x+51 | 244* / 56 / 2x+56 | 243* / 53 / 2x+53 | 247* / 55 / 2x+55 | 247* / 53 / 2x+53 | 251* / 55 / 2x+55 | 252* / 54 / 2x+54 | 257* / 57 / 2x+57 | 258* / 56 / 2x+56 | 259* / 55 / 2x+55 | 260* / 54 / 2x+54 |
| +3 | -3 | +5 | -3 | +5 | -3 | +2 | -2 | +2 | -1 | +3 | -1 | -1 | -1 |

[50]-1 | +1

© Darden H. Dickson 2007

Δαρδεν

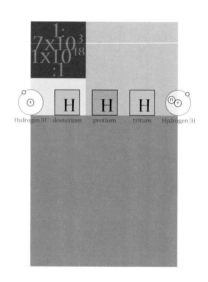

# MEMORY WORK SHEET

11

# Table of Proton Numbers

| THE NUMBER OF PROTONS | SYMBOL | NAME | NUCLEAR MASS NUMBER |
|---|---|---|---|
| 1 | H | Hydrogen | 1.00794 |
| 2 | He | Helium | 4.00260 |
| 3 | Li | Lithium | 6.941 |
| 4 | Be | Beryllium | 9.01218 |
| 5 | B | Boron | 10.811 |
| 6 | C | Carbon | 12.011 |
| 7 | N | Nitrogen | 14.0067 |
| 8 | O | Oxygen | 15.9994 |
| 9 | F | Fluorine | 18.9984 |
| 10 | Ne | Neon | 20.1797 |
| 11 | Na | Sodium | 22.9898 |
| 12 | Mg | Magnesium | 24.3050 |
| 13 | Al | Aluminum | 26.9815 |
| 14 | Si | Silicon | 28.0855 |
| 15 | P | Phosphorus | 30.9738 |
| 16 | S | Sulfur | 32.066 |
| 17 | Cl | Chlorine | 35.4527 |
| 18 | Ar | Argon | 39.948 |
| 19 | K | Potassium | 39.0983 |
| 20 | Ca | Calcium | 40.078 |
| 21 | Sc | Scandium | 44.9559 |
| 22 | Ti | Titanium | 47.88 |
| 23 | V | Vanadium | 50.9415 |
| 24 | Cr | Chromium | 51.9961 |
| 25 | Mn | Manganese | 54.9381 |
| 26 | Fe | Iron | 55.847 |
| 27 | Co | Cobolt | 58.9332 |
| 28 | Ni | Nickel | 58.69 |
| 29 | Cu | Copper | 63.546 |
| 30 | Zn | Zinc | 65.39 |
| 31 | Ga | Gallium | 69.723 |
| 32 | Ge | Germanium | 72.61 |
| 33 | As | Arsenic | 74.9216 |
| 34 | Se | Selenium | 78.96 |
| 35 | Br | Bromine | 79.904 |
| 36 | Kr | Krypton | 83.80 |
| 37 | Rb | Rubidium | 85.4678 |
| 38 | Sr | Strontium | 87.62 |

## Table of Proton Numbers

| THE NUMBER OF PROTONS | SYMBOL | NAME | NUCLEAR MASS NUMBER |
|---|---|---|---|
| 39 | Y | Yttrium | 88.9059 |
| 40 | Zr | Zirconium | 91.224 |
| 41 | Nb | Niobium | 92.9064 |
| 42 | Mo | Molybdenum | 95.94 |
| 43 | Tc | Technicium | 98 |
| 44 | Ru | Ruthenium | 101.07 |
| 45 | Rh | Rhodium | 102.906 |
| 46 | Pd | Palladium | 106.42 |
| 47 | Ag | Silver | 107.868 |
| 48 | Cd | Cadmium | 112.411 |
| 49 | In | Indium | 114.818 |
| 50 | Sn | Tin | 118.710 |
| 51 | Sb | Antimony | 121.75 |
| 52 | Te | Tellurium | 127.60 |
| 53 | I | Iodine | 126.904 |
| 54 | Xe | Xenon | 131.29 |
| 55 | Cs | Cesium | 132.905 |
| 56 | Ba | Barium | 137.327 |
| 57 | La * | Lanthanum | 138.906 |
| 58 | Ce * | Cerium | 140.115 |
| 59 | Pr * | Praseodymium | 140.908 |
| 60 | Nd * | Neodymium | 144.24 |
| 61 | Pm * | Promethium | 145 |
| 62 | Sm * | Samarium | 150.36 |
| 63 | Eu * | Europium | 151.965 |
| 64 | Gd * | Gadolinium | 157.25 |
| 65 | Tb * | Terbium | 158.925 |
| 66 | Dy * | Dysprosium | 162.50 |
| 67 | Ho * | Holmium | 164.930 |
| 68 | Er * | Erbium | 167.26 |
| 69 | Tm * | Thulium | 168.934 |
| 70 | Yb * | Ytterbium | 173.04 |
| 71 | Lu * | Lutetium | 174.967 |
| 72 | Hf | Hafnium | 178.49 |
| 73 | Ta | Tantalum | 180.948 |
| 74 | W | Tungsten | 183.85 |
| 75 | Re | Rhenium | 186.207 |
| 76 | Os | Osmium | 190.23 |

## Table of Proton Numbers

| THE NUMBER OF PROTONS | SYMBOL | NAME | NUCLEAR MASS NUMBER |
|---|---|---|---|
| 77 | Ir | Iridium | 192.22 |
| 78 | Pt | Platinum | 195.08 |
| 79 | Au | Gold | 196.967 |
| 80 | Hg | Mercury | 200.59 |
| 81 | Tl | Thallium | 204.383 |
| 82 | Pb | Lead | 207.2 |
| 83 | Bi | Bismuth | 208.980 |
| 84 | Po | Polonium | 209 |
| 85 | At | Astatine | 210 |
| 86 | Rn | Radon | 222 |
| 87 | Fr | Francium | 223 |
| 88 | Ra | Radium | 226.025 |
| 89 | Ac * | Actinium | 227.028 |
| 90 | Th * | Thorium | 232.038 |
| 91 | Pa * | Protactinium | 231.036 |
| 92 | U * | Uranium | 238.029 |
| 93 | Np * | Neptunium | 237.048 |
| 94 | Pu * | Plutonium | 244 |
| 95 | Am * | Americium | 243 |
| 96 | Cm * | Curium | 247 |
| 97 | Bk * | Berkelium | 247 |
| 98 | Cf * | Californium | 251 |
| 99 | Es * | Einsteninium | 252 |
| 100 | Fm * | Fermium | 257 |
| 101 | Md * | Mendelevium | 258 |
| 102 | No * | Nobelium | 259 |
| 103 | Lr * | Lawrencium | 260 |
| 104 | Rf | Rutherfordium | 261 |
| 105 | Db | Dubnium | 262 |
| 106 | Sg | Seaborgium | |
| 107 | Bh | Bohrium | |
| 108 | Hs | Hassium | |
| 109 | Mt | Meitnerium | |
| 110 | Uun | Ununnilium | |
| 111 | Uuu | Unununium | |
| 112 | Uub | Ununbium | |

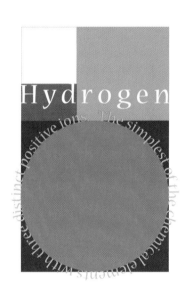

# Hydrogen

distinct positive ions. "The simplest of the chemical elements, with three

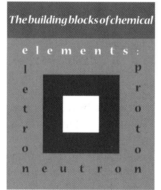

The building blocks of chemical

e l e m e n t s :

l         p
e         r
t         o
r         t
o         o
n  n e u t r o n